I0605340

THE PLANETS OF OUR SOLAR SYSTEM

JUPITER

by Arnold Ringstad

BrightPoint Press

San Diego, CA

an imprint of ReferencePoint Press, Inc.
Printed in the United States

For more information, contact:
BrightPoint Press
PO Box 27779
San Diego, CA 92198
www.BrightPointPress.com

LIBRARY OF CONGRESS CATALOGING-IN-PUBLICATION DATA

Name: Ringstad, Arnold, author.
Title: Jupiter / by Arnold Ringstad.
Description: San Diego, CA: ReferencePoint Press, 2026 | Series: The planets of our solar system | Audience: Grade 7 to 9 | Includes bibliographical references and index.
Identifiers: ISBN: 9781678211622 (hardcover) | ISBN: 9781678211639 (eBook)
The complete Library of Congress record is available at www.loc.gov.

CONTENTS

AT A GLANCE

- Jupiter is the largest planet in the solar system. It is sometimes called the king of the planets.
- Jupiter is a gas giant. This type of planet is made up of gases and has no solid surface.
- Jupiter's thick atmosphere has colorful swirling clouds. It is made up mainly of hydrogen and helium.
- Jupiter is visible to the naked eye in the night sky, so people have been observing it since ancient times.
- The invention of the telescope let astronomers learn much more about Jupiter. Galileo Galilei used one to discover the planet's four largest moons.
- The *Pioneer* and *Voyager* probes flew past Jupiter. Later, the *Galileo* and *Juno* probes entered orbit around Jupiter and studied the planet for years.

- Orbiting Jupiter are ninety-five moons and a thin set of rings.

- Jupiter's four largest moons are Ganymede, Callisto, Io, and Europa. They are known as the Galilean moons.

- Europa has an icy shell covering an ocean of liquid water. Scientists believe it may have the conditions needed for life. Future probes, such as *Europa Clipper*, will gather more data about the moon.

JUNO AT JUPITER

A long journey was coming to an end. The *Juno* spacecraft had left Earth in 2011. Then it drifted through space for 5 years. *Juno* traveled more than 1.7 billion miles (2.7 billion km). In July 2016, it finally approached Jupiter.

Scientists on Earth watched closely as *Juno* neared the giant planet. A team at the Jet Propulsion Laboratory (JPL) in California monitored the mission. Rows of scientists

This illustration shows *Juno* passing over Jupiter.

Technicians fuel *Juno* before its mission.

sat at computer screens. They studied their **data** for any problems.

Engineers designed *Juno* to go into **orbit** around Jupiter. To do that, it needed to slow down. Otherwise it would fly right past the planet. The spacecraft's rocket engine started right on time. The engine fired in the opposite direction of the spacecraft's motion. This slowed *Juno* down. The engine needed to fire for 35 minutes for *Juno* to reach the correct speed. The scientists nervously waited as the minutes passed.

Finally, the signal arrived from the spacecraft. The engine had shut down on time. *Juno* was now in orbit around Jupiter. The control room erupted in cheers and applause.

Reaching Jupiter was a big success. But it was just the beginning of *Juno*'s true mission. The spacecraft was built to study Jupiter. It carried many scientific instruments. Some analyzed Jupiter's

***Juno* took detailed observations of Jupiter's clouds.**

powerful gravity. Others peered into the planet's swirling clouds. And some provided data about Jupiter's rings and moons.

A GIANT WORLD

Jupiter is the solar system's largest planet. In fact, it has more mass than all the other planets combined. It is a type of world known as a gas giant. It has no solid surface beneath its colorful clouds. This means spacecraft cannot land there.

People have learned a lot about Jupiter using telescopes on Earth. Spacecraft have also studied this planet up close. Jupiter's beautiful clouds and many moons make it one of the solar system's most fascinating planets.

CHAPTER ONE

KING OF THE PLANETS

Jupiter is the fifth planet from the Sun. It is located in the outer solar system. The **asteroid** belt separates this region from the inner solar system. The outer solar system contains four planets. Besides Jupiter, they include Saturn, Uranus, and Neptune. These four worlds are giant planets.

Jupiter is sometimes called the king of the planets. This is because of its

This montage of the solar system uses images taken by spacecraft. The lower four planets are the solar system's giant planets.

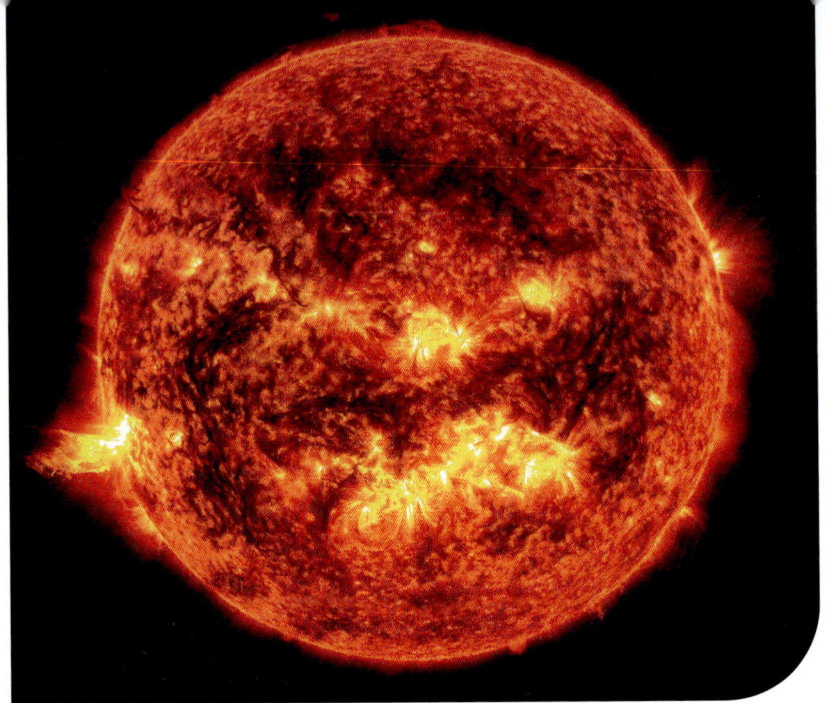

The Sun is the only object in the solar system that is more massive than Jupiter.

enormous size. The planet's diameter is more than ten times that of Earth. More than 1,300 Earths could fit inside Jupiter. If Earth were a grape, Jupiter would be a basketball. The planet's mass may be even more impressive. Jupiter has more than twice the mass of all the other planets put together.

That great mass means Jupiter has strong gravity. This gravity affects the paths of objects that come near. Jupiter may even protect Earth from dangerous asteroids. Writing for Space.com, author Charles Q. Choi explains, “Its enormous gravity can suck in and absorb smaller objects . . . or propel them out of the solar system entirely.”[1]

Jupiter spins on its **axis** about every 10 hours. This means it has the fastest day of any planet. But its year is very long. This is because of its great distance from the Sun. On average, Jupiter is about 484 million miles (778 million km) from the Sun. It takes a long time to complete an orbit. Each orbit takes about 12 Earth years.

A planet’s tilt on its axis causes differing amounts of sunlight to fall on different parts of the planet as it orbits. This is what creates seasons. The bigger the tilt, the stronger this effect is. Earth tilts at 23.5 degrees, creating significantly different seasons. Jupiter's tilt is just 3 degrees. This means its seasons are very similar.

JUPITER'S ATMOSPHERE AND LAYERS

Jupiter has a thick, dense atmosphere. This layer of gases is made up mostly of hydrogen, along with some helium. There are also much smaller amounts of chemicals such as ammonia, methane, and water.

Jupiter's clouds are visible from space. They form white and red bands that mix and swirl. The white bands are colder and are known as zones. The red bands are known as belts. Clouds of water interact with the bands, too. Jupiter's atmosphere is in constant motion.

Besides these bands, Jupiter's atmosphere has other features. The most famous is known as the Great Red Spot. This reddish, oval-shaped storm has existed on Jupiter for hundreds of years. It is larger

The Great Red Spot changes in size, shape, and color over time.

than Earth. The first recorded observation of the Great Red Spot was in the 1830s. The storm still blows today. However, it is shrinking over time. Someday it may disappear completely.

Scientists cannot see what lies below Jupiter's atmosphere. But they can study the planet to guess at what might be there. Deeper into Jupiter's clouds, the pressures and temperatures rise. Hydrogen goes from a gas to a liquid. At even greater depths, it becomes a metallic solid. At the planet's center, scientists believe there is a solid core that is about the size of Earth.

MOONS, RINGS, AND ASTEROIDS

Jupiter has ninety-five confirmed moons. This is the second most behind Saturn.

The four largest are Ganymede, Callisto, Io, and Europa. The rest of the moons are much smaller. In fact, these four large bodies contain more than 99.9 percent of the ninety-five moons' total mass.

Jupiter is surrounded by a thin set of rings made up of dust. These were

Comet Crash

In 1993, **astronomers** discovered a **comet** near Jupiter. They were surprised to learn it would crash into the planet the next year. In July 1994, spacecraft and telescopes closely watched Jupiter. The comet had already been ripped to pieces, and the chunks hit Jupiter over several days. They left dark marks that eventually vanished in the clouds. Studying the impacts helped scientists learn more about Jupiter's atmosphere.

This illustration depicts the two groups of Trojans in Jupiter's orbit.

unknown to scientists until 1979. In that year, the *Voyager 1* spacecraft saw them as it flew past the planet. Later research showed how the rings formed. Space rocks crashed into Jupiter's small moons. This sent clouds of dust into space. Jupiter's gravity pulled this dust into rings around the planet.

The king of the planets is joined in its orbit by two groups of asteroids. One swarm is ahead of Jupiter, and another follows behind. They are held in place by the combination of Jupiter's gravity and the Sun's gravity. These asteroids are known as the Trojans. Scientists estimate there are more than 250,000 Trojans of Jupiter larger than 0.62 miles (1 km) in diameter.

The asteroid belt lies between the orbital paths of Jupiter and Mars. This wide region consists of more than 1 million asteroids larger than 0.6 miles (1 km) in diameter. It also contains millions of smaller asteroids. Jupiter's gravity plays a big role in keeping the asteroid belt in place. Without Jupiter, many of the asteroids would have become part of the other planets.

CHAPTER TWO

OBSERVING JUPITER

Jupiter is a bright object in the night sky. This means people have been able to see it since prehistoric times. The ancient Babylonians tracked the planet's motion across the sky. Astronomers in ancient China also recorded observations.

The planet's name came from the ancient Romans. They called this bright object Jupiter, after the king of the gods.

This Roman statue of Jupiter was constructed in the first century CE. It now stands in a museum in Russia.

Galileo Galilei demonstrates his telescope to the people of Venice, Italy.

Jupiter was a god of sky and thunder. He was a version of the earlier Greek god Zeus.

JUPITER THROUGH TELESCOPES

Craftsmen in the Netherlands invented telescopes in 1608. At first, people used these devices to see distant things on Earth. But soon scientists began pointing them upward. They studied the night sky and made new discoveries. One of these people was Italian astronomer Galileo Galilei.

Galileo built his own telescope and observed Jupiter. He saw bright lights near the planet. Galileo published his discovery in 1610. He wrote, “Our own eyes show us four stars which wander around Jupiter as does the Moon around the Earth.”[2] Galileo realized that these were moons circling Jupiter.

In 1614, German astronomer Simon Marius published his own findings. He claimed he discovered the moons first. People at the time thought he was copying Galileo. But historians now think the two astronomers discovered the

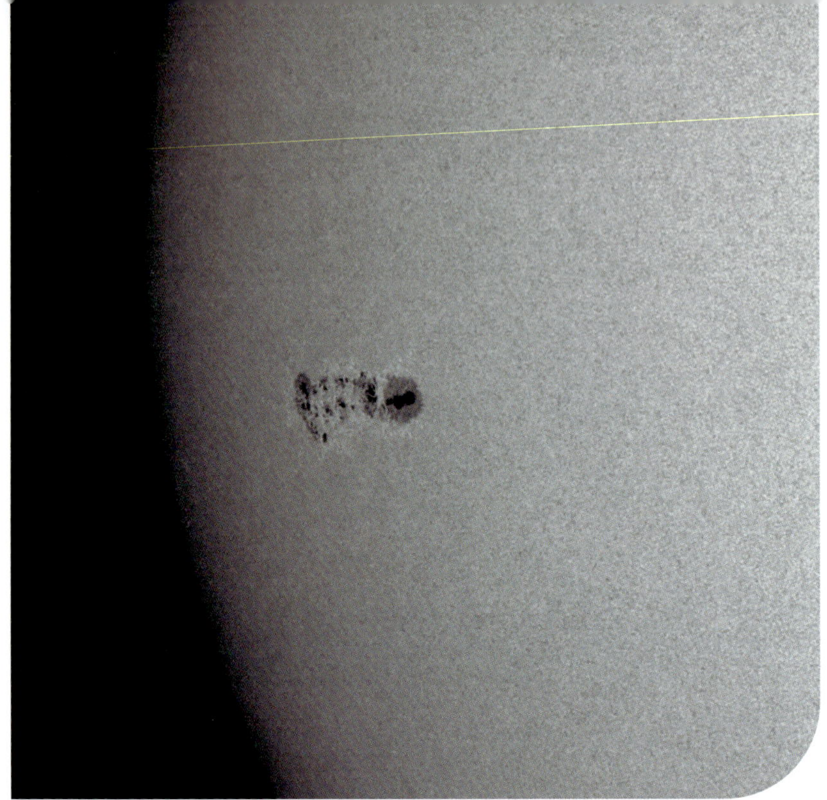

Samuel Heinrich Schwabe made many observations of sunspots. These dark spots occasionally appear on the Sun's surface.

moons independently. Marius had taken longer to publish his original observations.

SEEING THE SPOT

Scientists improved telescopes over time. This gave them better views of the night sky. It led to new discoveries about the planets, including Jupiter.

One of these discoveries came from German astronomer Samuel Heinrich Schwabe. He worked in the early 1800s. He is best known for his many years of observations of the Sun. But he also looked

Io and the Speed of Light

Danish astronomer Ole Rømer worked in the 1670s. He watched as Jupiter's moons were eclipsed, traveling behind the planet and then reappearing. The timing of eclipses seemed to vary based on Earth's position. He wondered if this was because light had a specific speed instead of being infinitely fast. The light of the eclipse took longer to reach Earth when the two planets were farther apart. Based on this, Rømer predicted the timing of an eclipse of Io in 1676. He was right, showing that the speed of light was finite.

at Jupiter. In 1831, he described and drew the Great Red Spot for the first time.

JUPITER'S RADIO WAVES

Astronomers use more than just telescopes to learn about the solar system. They can also use antennas to collect radio waves that come from space. In 1955, astronomers Bernard Burke and Kenneth Franklin did this. They placed their antennas near Washington, DC. They soon noticed a strong radio signal arriving about once per day.

Burke and Franklin worked to explain the signal. They tried to figure out what objects in the night sky could be causing it. Finally they realized the signal came when the antennas pointed at Jupiter. Franklin later

Scientists use radio telescopes to study distant sources of radio waves. This radio telescope was the largest in the world when it was completed in 2016.

wrote, "No other object could satisfy the data: the source of the . . . **radiation** was definitely associated with Jupiter!"[3]

Jupiter was giving off radio signals. The pattern of signals matched Jupiter's rotation. This helped them figure out how long the planet takes to rotate on its axis.

CHAPTER THREE

A GAS GIANT UP CLOSE

In the mid-1900s, advancing rocket technology made spaceflight possible. The first spacecraft launched into orbit around Earth in 1957. In the following decades, people sent **probes** out into the solar system to study planets and moons. Several of these missions explored Jupiter.

The twin probes *Pioneer 10* and *Pioneer 11* were the first to reach the king of the planets. The National Aeronautics

***Pioneer 10* was launched into space by an Atlas-Centaur rocket.**

UNITED
STATES

and Space Administration (NASA) is the US space agency. It built and launched the probes. *Pioneer 10* blasted off in March 1972. *Pioneer 11* followed in April 1973.

Jupiter is far away, and it takes many months to get there. *Pioneer 10* reached

Technicians work on *Pioneer 10* before its launch.

Jupiter in December 1973. It took close-up photos of the planet. The probe also gathered data about Jupiter's radiation and **magnetic field**. *Pioneer 11*'s main destination was Saturn, but it flew past Jupiter on the way. It captured more images of the planet and its moons.

Going into orbit around a planet means slowing down. This requires a lot of fuel. The lightweight *Pioneer* probes were not designed for this. They simply studied Jupiter while zooming past the planet. Then they continued traveling into deep space.

THE *VOYAGERS*

The next spacecraft to visit Jupiter were *Voyager 1* and *Voyager 2*. These twin probes launched in 1977. At the time, the

The *Voyager* probes have large dish-shaped antennas that can send and receive signals across great distances.

outer planets were lined up in a special way. The probes could fly from one to the next much faster than they normally could. This path through the solar system is sometimes called the grand tour.

The two probes reached Jupiter in 1979. They captured the best images yet of the planet. They learned more about the Great Red Spot. And they discovered a thin ring around Jupiter.

The *Voyager* probes also examined the planet's moons. They discovered volcanoes on Io and icy cracks on Europa. They also made an accurate measurement of Ganymede's size. They confirmed it is the biggest moon in the solar system.

Linda Morabito worked on the *Voyager* team. She studied images the probes sent back. Morabito helped discover the volcanoes on Io. She recalls the moon's strange appearance. She says, "I will never forget it. The colors of Io, caused by sulfur dioxide at various temperatures, range from black to orange to yellow to blue. The way scientists described it and the way we described it caught on very quickly: it looks like a moldy pizza!"[4]

As with *Pioneer 10* and *11*, the *Voyager* mission was a flyby. The spacecraft did not go into orbit around Jupiter. Instead, they continued on to Saturn. *Voyager 1* studied Saturn and its moon Titan. Then it left on a path out of the solar system. *Voyager 2* kept going to Uranus and finally to Neptune. The spacecraft are now billions of miles away. In 2025, they still had enough power to communicate with Earth.

GALILEO

The *Galileo* mission provided a wealth of information about Jupiter. This probe launched from Earth in 1989. It took a long path to Jupiter. *Galileo* first flew past Venus

THE *VOYAGER* PROBES' TOUR OF THE SOLAR SYSTEM

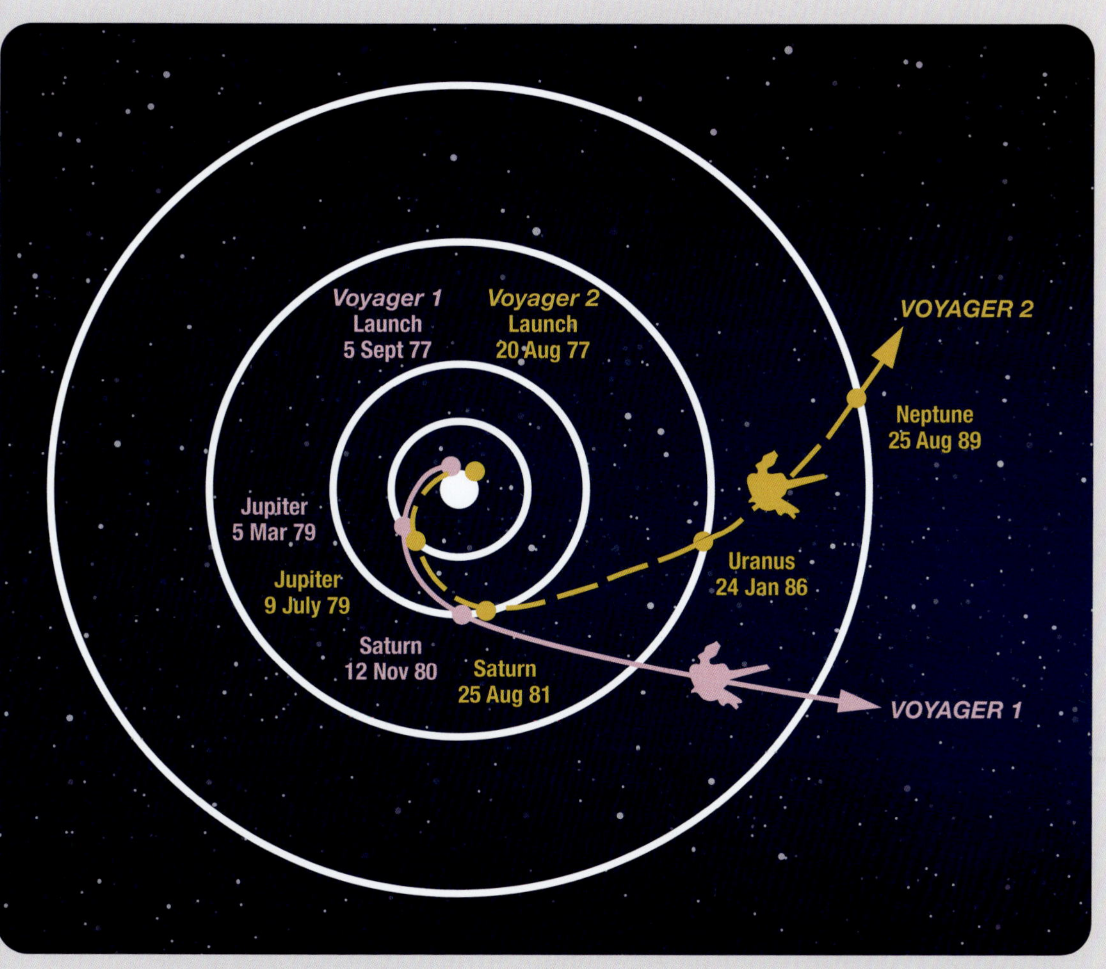

Voyager 1 **launched after** ***Voyager 2*****, but it took a shorter path through the solar system.** ***Voyager 1*** **is the most distant human-made object from Earth.**

and Earth. It used the planets' gravity to change its speed and direction. It finally reached Jupiter in late 1995. It used its engine to slow down and enter orbit around the planet.

This NASA illustration shows *Galileo* passing closely above Io with Jupiter in the background.

As *Galileo* approached Jupiter, it released a smaller probe. This tough, tiny probe flew straight into the planet. It had to survive amazing speeds, temperatures, and pressure. It used a parachute to slow its fall through Jupiter's clouds. Eventually Jupiter's harsh conditions destroyed the probe. But before then, it sent back valuable data on the planet's atmosphere.

Hubble Looks at Jupiter

Spacecraft that visit Jupiter are able to get the best views of the planet. But telescopes orbiting Earth can get great images as well. The Hubble Space Telescope has been in space since 1990. It can capture high-quality photos of Jupiter. It is especially useful for studying planets over many years. Its images can show long-term changes in Jupiter's atmosphere.

Galileo spent 5 years orbiting Jupiter. It took amazing photos of the planet's swirling clouds. Some instruments gathered data on Jupiter's radiation and magnetic field. Others examined the elements that make up the atmosphere. *Galileo* also made close approaches to Jupiter's moons. In 2001, it passed only about 112 miles (181 km) above Io.

In 2003, *Galileo*'s mission came to an end. The spacecraft was running out of fuel. Scientists did not want to risk the probe hitting Europa. They had discovered this moon likely had an ocean under its icy surface. Scientists feared that *Galileo* could disturb this ocean's environment. So, they commanded *Galileo* to fly into

Jupiter's atmosphere. The probe burned up as it fell.

JUNO'S DISCOVERIES

The *Juno* probe arrived at Jupiter in 2016. Like *Galileo*, it is an orbiter. This would let it study the gas giant for several years.

At Jupiter, *Juno*'s solar arrays generated up to 500 watts of power. This was enough energy for all of the spacecraft's operations.

Juno has one key difference from earlier probes. *Pioneer*, *Voyager*, and *Galileo* used radioactive power sources to create electricity. *Juno* uses solar panels instead. But Jupiter is very far from the Sun. This means the Sun's light is dimmer than it is near Earth. *Juno* needs huge solar panels to generate enough power. Its three solar arrays are each 30 feet (9 m) long.

Juno's instruments can peer deep into Jupiter's atmosphere. It has learned more about the planet's weather, including its storms and lightning. It has helped scientists understand the planet's core. Like past probes, *Juno* has also spent time studying Jupiter's moons. It has taken sharp photos of Ganymede, Europa, and Io.

Pictures taken by *Juno* revealed complex storms in Jupiter's northern regions.

***Europa Clipper* flies by Europa in this illustration by NASA.**

The original plan was for *Juno*'s mission to last until 2021. But when that time came, the spacecraft was still working well. NASA extended the mission to late 2025. Scientist Scott Bolton explains, "Since its first orbit in 2016, Juno has delivered one revelation after another about the inner workings of this massive gas giant. With the extended mission, we will . . . [reach] beyond the planet to explore Jupiter's ring system and Galilean satellites."[5]

FUTURE MISSIONS

In 2025, advanced new probes were on their way to Jupiter. One was the *Jupiter Icy Moons Explorer* (*Juice*). *Juice* launched in 2023. It was due to reach Jupiter in 2031. The probe's mission was to study the icy oceans of Europa, Ganymede, and Callisto.

Europa Clipper launched in 2024. It was taking a more direct route to Jupiter than *Juice*. This meant it would arrive first, in 2030. *Europa Clipper* will focus mainly on the fascinating moon Europa. Its instruments will study whether life could exist in the moon's ocean.

JUPITER'S MOONS

Jupiter has ninety-five officially recognized moons. Four of Jupiter's moons are by far the largest. These are Ganymede, Callisto, Io, and Europa. The other ninety-one moons are tiny by comparison.

These four large moons were discovered by Galileo. Together they are known as the Galilean moons. However, the Italian astronomer did not name them.

Many of Ganymede's craters are visible in this *Juno* image of the moon.

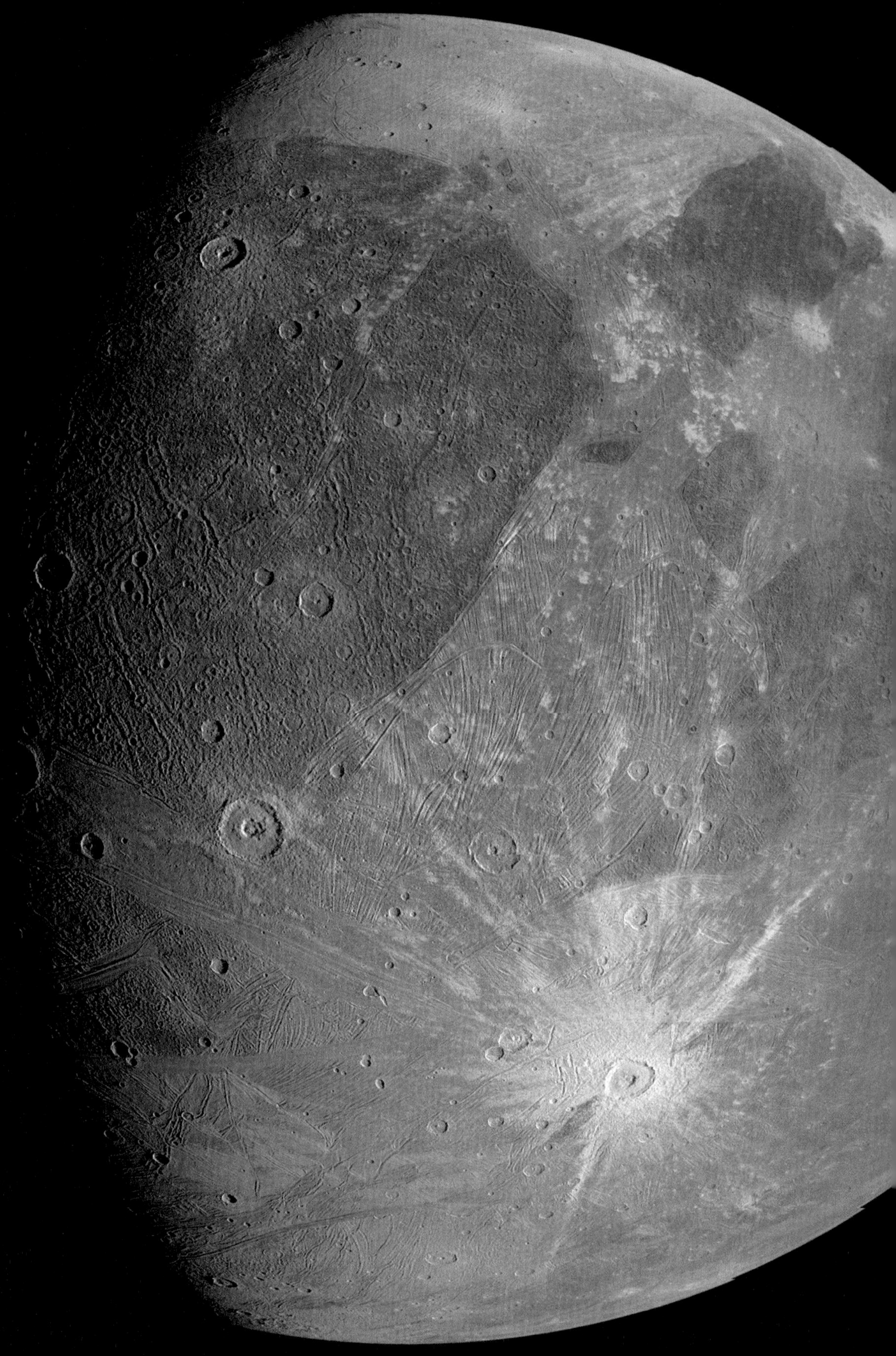

The names came from German astronomer Simon Marius.

GANYMEDE

Ganymede is Jupiter's largest moon. It is also the biggest moon in the solar system. It is even a bit larger than the planet Mercury.

On Earth, auroras occur when bursts of energy from the Sun strike the atmosphere.

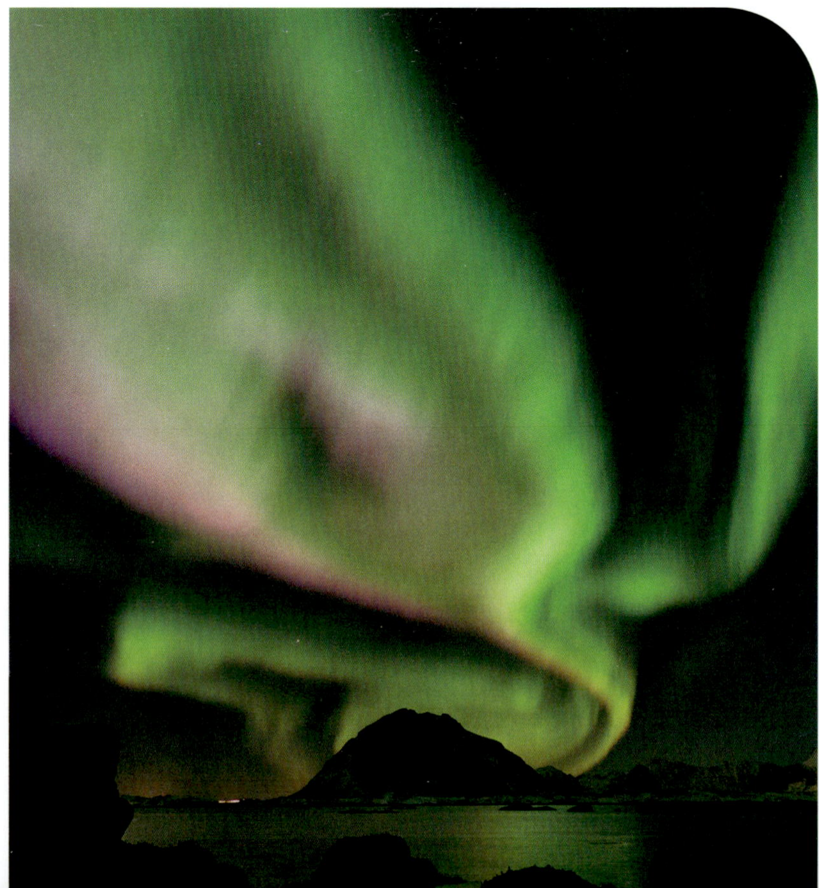

Close-up photos from probes reveal craters, dark areas, and lighter regions on Ganymede. Scientists suspect the moon may have a salty ocean under the surface. They have discovered that Ganymede is the only moon with a magnetic field. As a result, it experiences brightly colored auroras like those on Earth.

CALLISTO

Callisto is the next-largest moon of Jupiter. It is covered with a huge number of craters. Bright spots on the moon's surface may be patches of ice.

The *Galileo* probe discovered evidence of a salty sea under Callisto's surface. However, later studies caused scientists to rethink this. The sea may be deeper

***Galileo* captured this image of craters on Callisto's surface.**

underground than originally thought. Or there may be no underground sea at all. Future probes may help unlock Callisto's secrets.

IO

Jupiter's third-largest moon is Io. This moon is best known for its many volcanoes. Its surface is always changing.

Io's volcanic activity is created by Jupiter's intense gravity. Jupiter tugs at Io's surface, creating heat within the moon. Underground melted material erupts to the surface in volcanoes. This creates lakes of lava and smooth surfaces of cooled rock.

Io is the most volcanically active object in the solar system.

Unlike with the other Galilean moons, there is no evidence of water on Io.

EUROPA

The last Galilean moon is Europa. It is slightly smaller than Earth's moon. Europa has a thick, icy shell. This surface is covered

The cracks on Europa's surface form due to the strong pull of Jupiter's gravity.

with reddish-brown cracks. There are few craters. Below the surface is a salty ocean. Scientists estimate it is up to 100 miles (160 km) deep. NASA's Cynthia Phillips explains, "We think that Europa has more water than all of Earth's oceans combined."[6]

Europa has some of the ingredients needed for life. There is liquid water in its oceans. It has chemicals that life needs, such as carbon and oxygen. And radiation from Jupiter provides energy that could help life grow. Scientists are studying the possibility of life deep under Europa's surface. *Europa Clipper* and other missions seek to learn more about this.

NASA scientist Lynnae Quick-Henderson describes what a person standing on Europa might experience:

It would be frigid, first of all. We'd have to have the heaviest winter coats that would ever exist. But we'd see mountains. These ridges would look . . . a lot like mountains do. And we might see a geyser erupting in the distance. We might see the icy particles and the water vapor spraying up over the surface of Europa.[7]

OTHER MOONS

Four other moons of Jupiter are known as the inner group. They include Metis, Adrastea, Amalthea, and Thebe. Each is less than 124 miles (200 km) across. They orbit close to Jupiter, moving in the same direction as the planet's rotation.

NASA's *New Horizons* spacecraft took these images of Jupiter's rings.

Their gravity helps hold Jupiter's thin rings in place.

Metis, Adrastea, and Thebe were discovered in images taken by the *Voyager* probes. But Amalthea was discovered before spaceflight was even possible. In 1892, American astronomer Edward Emerson Barnard saw Amalthea through a telescope. The *Voyager* and *Galileo* probes later took observations of this moon.

The remaining moons are known as Jupiter's irregular satellites. These moons are organized into different groups. Moons within a group share similar orbits. They may have been created by asteroids

Collision Warning

In 2017, a team led by American astronomer Scott Sheppard discovered Jupiter's moon Valetudo. It's a tiny body less than 0.6 miles (1 km) across. This makes it the smallest known moon of Jupiter. Sheppard's team found that Valetudo was headed for a possible collision. It orbits in the opposite direction of nearby moons, and their orbital paths cross each other. Someday Valetudo may smash into another moon.

crashing together after being captured by Jupiter's gravity.

One example is the Himalia group. Its name comes from the moon Himalia, which is about 53 miles (85 km) across. Its orbit is similar to those of Leda, Lysithea, and Elara. Other groups include the Carme group and the Ananke group.

Jupiter's many moons are just one reason for the planet's popularity. The gas giant's swirling clouds make it a favorite of astronomers and everyday people alike. People can even observe Jupiter's largest moons through a backyard telescope. The king of the planets remains one of the solar system's most fascinating worlds.

GLOSSARY

asteroid

a large rock moving through space

astronomers

scientists who study space

axis

an imaginary line running through the center of a sphere

comet

a space object composed of ice and dust that leaves behind a long tail of material due to the Sun's energy

data

information gathered for a purpose

magnetic field

the area around a magnetic material within which the force of magnetism works

orbit

the path that one object takes around another object in space

probes

devices sent to space that gather information

radiation

the emission of energy in waves or particles

SOURCE NOTES

CHAPTER ONE: KING OF THE PLANETS

1. Charles Q. Choi, "Jupiter: A Guide to the Largest Planet in the Solar System," *Space.com*, March 18, 2023. www.space.com.

CHAPTER TWO: OBSERVING JUPITER

2. Quoted in "1610 Activity: Observing the Moons of Jupiter," *Everyday Cosmology*, n.d. https://cosmology.carnegiescience.edu.

3. K.L. Franklin, "An Account of the Discovery of Jupiter as a Radio Source," *The Astronomical Journal*, March 1959, pp. 37–39.

CHAPTER THREE: A GAS GIANT UP CLOSE

4. Quoted in Iain Todd, "Interview with *Voyager* Scientist Linda Morabito," *BBC Sky at Night Magazine*, July 6, 2020. www.skyatnightmagazine.com.

5. Quoted in "NASA's *Juno* Mission Expands Into the Future," *Mission Juno*, January 21, 2021. www.missionjuno.swri.edu.

CHAPTER FOUR: JUPITER'S MOONS

6. Quoted in Lulu Miller, "Kids Ask a NASA Scientist About the Mission of Jupiter's Moon Europa," *NPR*, October 18, 2024. www.npr.org.

7. Quoted in "*Europa Clipper*'s Voyage to Jupiter's Ocean Moon," *NASA*, October 1, 2024. www.nasa.gov.

FOR FURTHER RESEARCH

BOOKS

Heather C. Morris, *Saturn*. BrightPoint Press, 2026.

Gail Radley, *The Space Encyclopedia*. Abdo Publishing, 2023.

Henrietta Toth, *Robotics in Space*. BrightPoint Press, 2023.

INTERNET SOURCES

"Hubble Tracks Jupiter's Stormy Weather," *NASA*, March 14, 2024. https://science.nasa.gov.

"Jupiter," *The Planetary Society*, 2025. https://planetary.org.

"Jupiter: A Giant Among Giants," *National Air and Space Museum*, n.d. https://airandspace.si.edu.

WEBSITES

NASA: Eyes on the Solar System
https://eyes.nasa.gov/apps/solar-system

This interactive website allows users to view a simulation of the solar system. It also offers basic facts and shows the planets' rings and the relative positions of moons.

NASA: Solar System Exploration
https://science.nasa.gov/solar-system

NASA's website about the exploration of the solar system features information about planets, moons, asteroids, and more. This website also provides helpful information for people interested in observing the night sky.

The Schools' Observatory
www.schoolsobservatory.org

The Schools' Observatory is an online educational resource that provides information about how students can observe the night sky and what they might learn in doing so. The website also has information for students interested in pursuing a career in astronomy.

INDEX

IMAGE CREDITS

Cover: © Kevin M. Gill/NASA
5: © Andrea Luck/NASA
7: © NASA
8: © NASA
10: © Kevin M. Gill/NASA
13: © NASA
14: © NASA
17: © NASA
20: © NASA
23: © Karasev Viktor/Shutterstock Images
24: © Oxford Science Archive/Heritage Images/The Print Collector/Alamy
26: © NASA
29: © Jeff Dai/Stocktrek Images, Inc./Alamy
31: © NASA
32: © NASA
34: © NASA
37 (diagram): © NASA
37 (background): © KK.Kickin/Shutterstock Images
38: © NASA
41: © NASA
43: © Kevin M. Gill/NASA
44: © NASA
47: © NASA
48: © Johannes Groll/Unsplash
50: © NASA
51: © NASA
52: © NASA
55: © NASA

ABOUT THE AUTHOR

Arnold Ringstad lives in Minnesota. He enjoys reading and writing about space exploration.